BASIC

INTRODUCTION

TO ASSAYING

BY

FIRE

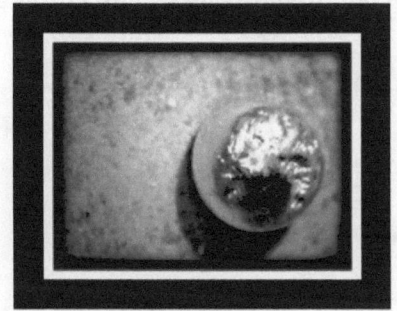

BY: N. E. KENDALL

Table of Contents;

The Table of contents is in the following order page numbers vary.

Introduction

The description of a Fire Assay for Gold and Silver

The use of 29.166 grams (1 Assay ton) of ore or concentrate to determine the amount of precious metals within one short ton, (2000 Lbs. North American) ton of material.

In the UK an Assay ton would be 32.23 grams, which would be representative of a Long Ton or 2,240 Lbs. the amount which bears the same ratio to a milligram as a short or long ton bears to a troy ounce. In other words, the number of milligrams of a particular metal found in a sample of this size gives the number of troy ounces contained in a short or long ton of ore.

Assay by fire is also the most elaborately accurate, but totally destructive process for precious metal determination. (It also be refers to the critical cupellation step that separates precious metal from lead along with the parting of silver from the precious metal bead.) If performed on bullion (high purity precious metal alloy) in accordance to international standards, the method can be accurate on gold metals to 1 part in 10,000.

If performed on ore materials using fusion followed by cupellation separation, detection may be in parts per billion. However, accuracy on ore material is typically limited to 3 to 5% of reported value. Although time consuming, the method is the accepted standard applied for valuing gold ore as well as gold and silver bullion at major refineries and gold mining companies.

In the bullion fire assay process, a sample from the article is wrapped in a lead foil with copper and silver. The wrapped sample is heated at 1850 F (temperature varies with exact method) in a cupel made of compressed bone ash or magnesium oxide powder. The Base metals oxidize and absorb into the cupel. The product of this cupellation (doré) is flattened and treated in a 25% nitric acid to remove silver. Precision weighing of metal content of samples and process controls (proofs) at each process stage is the basis of the extreme method precision. Reputable North

American bullion assayers conform closely to ASTM method E1335-04e1. Only bullion methods validated and traceable to accepted international standards obtain genuine accuracies of 1 part in 10,000.

Principle Assay Materials and tools

The Primary tools for performing a fire assay are as follows. A more extensive explanation is provided later in this book.

- A Melting Furnace
- Crucibles (Clay-Silica)
- Cupels (sized to absorb all the lead of the prill)
- Cupelling furnace or Kiln. Lifting and or pouring tongs
- Fluxing reagents, Nitric Acid
- Weight and measurement devices, comparator or appropriate balances
- Various Labware, such as a small beaker, tweezers etc. And last but not least all safety gear

Fluxes and their use:

Flux is generally a compound mixture suitable for collecting all the precious metals within the material to be assayed.

Combinations of these assay chemicals are about as various as chicken soup, however all accomplishing the same thing.

In general good starter for a common Assay Flux normally consists of about 69% Litharge (PbO) Lead oxide, along with 23.3% Soda ash, Borax 3.5%, Silica 2.2%, and Wheat Flour 2%. The lead oxide during furnacing or the fusion stage will be converted (reduced) to a solid lead precipitant, raining down thru the flux mixture fusing and collecting the precious metals as they fall by gravity to the bottom of the crucible. The Wheat Flour for carbon (A Reducer), needed is generally added as well generally mixed in last as in a few cases is not needed for reduction.

Melting, Smelting and refining fluxes differ as they normally contain Niters to oxidize the base metals.

The back pages of this book provide the extensive variances in fluxing which may be applied for tailoring to particular types of ores

Safety is without doubt the most important factor of a successful Assay.

MSHA/OSHA, your state agencies along with company policies and guidelines need to be followed to the letter, when dealing with poisonous mixtures, hot metals, chemicals and furnaces.

All, and any more precautions should be taken when working on or around these procedures. PPE is of the Utmost importance!

Always Read the MSDS sheets for any compounds you use,

Know who to call and what to do in case of an accident or emergency.

Sample Preparation, Fusion, and Pouring of the Lead Prill

Preparation of a sample is very important when it comes to getting an accurate fire assay, as whatever you prepare is an indication of the general lot it represents.

Splitting of the sample is probably the most important stage of an assay. Material that is not properly split cannot be accurately reproduced.

Example; Receipt of 2 lbs. of material is received, crushed and pulverized out of the 2 lbs. you need 29.166 grams (*one assay ton*) for your test. If taking only that amount without homogenization of the total sample, the assay will not be accurate, as it will represent only what you have taken and not the entire lot. Cupellation alone can only remove a limited quantity of impurities from a sample. Fire assay, as applied to ores, concentrates or less pure metals, adds a fusion or scorification step before cupellation.

Fusion and Pouring of the Lead Prill

Fusion Method is a melt (typically at 2200° -2300° F) in a dry chemical flux designed to precipitate lead and precious metals from the ore sample into lead button. Silicates, carbonates, and other non-precious impurities reject into a glassy slag. The lead button product is typically cupelled to further concentrate the product to pure precious metals, but selected instrument method are able to directly analyze precious metals within the lead button.

The Methods and details for various fire assay procedures vary extensively, but concentration and separation chemistry do today typically comply with traditions set by Bugby or Shepard & Dietrich in the early 20th century. Method advancements since that time have mostly automated the handling of sample materials and final computations, i.e., instrument computations rather than the physical gold weighing. Even these instructions and methods are mostly twists of traditions that were detailed in 1556.

Some variation from skills taught in the modern standards of fire assay methods should be looked upon very cautiously. The normal or standard traditions have an extensive history of reliability; "special" new methods may associate with reduced assay accuracy.

There are several methods for determining the amounts of precious metals in an assay bead. The two methods of determination of gold and silver (Au &Ag), being shown are a quicker method of determining values and losses in a production scenario.

First will be a Fusion method whereby the ore/concentrates will be fused with a lead precipitate.

(Explanation of precipitation) Carbon converts or reduces lead oxide into pure metal and has a raining effect collecting the minute particles of gold with gravity carrying the fused lead and precious metals to the bottom of the crucible.

Whereby the lead precipitate has fused and collected the precious metals into the bottom of the crucible, then after the allotted time where the Ferro metals and most other base metals have separated into the Glass (Flux) are poured into a conical mold to cool.

The glass is then broken or knocked off of the lead Prill pounding it into a cube placing both the lead and cupel into a high temperature oven until the lead has all oxidized (returned to PbO state) and absorbed into the cupel leaving only a precious metal bead.

NOTE: *In some cases Cupellation alone can only remove a limited quantity of impurities from a sample. Fire assay, as applied to ores, concentrates or less pure metals, adds a fusion or scorification step before cupellation. This is done in General to pre-oxidize some of the lead and impurities to create a smaller lead prill.*

Basic introduction to Assaying by Fire (History)

The following procedures are an introductory level instruction with the objective of passing along information and instruction by use of the art of the Fire assay. A pyrometallurgical procedure of analyzing the presence and amounts of Precious metals generally for the Gold and Silver contents for assay, usually expressed in Tons for a Short ton, US (2000 Lbs.) or Tonnes for a Long Ton, British, (2204 Lbs.).

Fire assaying is a procedure which has been thought to have been around for about 5000 years, the 15th century however, is when the first complete metallurgical work and was recorded by a German Scientist. A time proven method, being the most accurate, dependable, method of analysis known for use of determining amounts of precious metal in ores and other materials.

Typically the reporting error can be viewed as being +/- 5% but is still the preferable and standard acceptable method for determining the values of Gold and Silver.

The standard fire assay in the US is normally started with 29.166 grams (1 Assay ton) of ore. The weight of 29.166 grams used in assaying, is primarily for convenience.

This weight bears the same relation to the milligram that a short ton of 2000 avoirdupois pounds does to the troy ounce, the weight in milligrams of precious metal obtained from an assay ton of ore gives comparatively the number of ounces to the ton.

Looking forward is a good example, that when an assay bead is weighed after Cupellation a bead that weighs one milligram (1mg) is equal to 1 troy ounce (OzT) per ton of material.

Basic introduction to Assaying by Fire (Weights and Measures)

Although the standard assay is measured in the 29.166 sample, this number can be multiplied many times then the bead weight divided back to its single value in order to obtain an accurate weight. (some scales available and useable in less than a designed assay lab may not be capable of weighing less than 1mg.) This method may at this point compensate for the smaller bead allowing an accurate weight by increasing its size and weight. An example is that the bead seems too small and won't indicate a weight using a scale that has a minimum of 1mg capability.

The next assay that may be performed an assayer, may prefer to double the weight of the sample producing a bead that can now be weighed, the weight is 1.1mg now calculate this weight divided by the number of assay ton samples to produce a weight of .55mg which would be the accurate weight of the bead indicating that there are .55 OzT or slightly over a half troy ounce per ton of precious metals.

One Pound and half pound assays can be done very easily by using 1Lb pound of material and taking the direct bead weight X 2000 which normally produces a bead weighing in the milligram range, dividing by 1000, then, dividing the grams by 31.1 to achieve a Troy Ounce weight.

This can be a costly method because of the costs of the larger crucibles and Cupels but is highly accurate when all proper quality control is practiced.

Basic introduction to Assaying by Fire (Tools and Materials)

All of the Basic tools that are needed to do an assay are critical since accuracy is a necessity and the fact that an assayer will work with very extreme temperatures and conditions. Therefore, contrary to what seems to be posted on the internet and some not so professional companies which offer assaying advice, are no shortcuts or cheap tools and materials.

One must remember when procuring products for assaying that one will find a wide array of producers, most manufacturers of these products are not located in the US, so when one wants to find a product from a distributor, you must know, test, or be familiar with the particular product before a decision is made to use this in production.

To start with, described below you will find some of the basic tools needed for assaying, note many are handmade since many of these tools have a tendency to be tailored for a particular application, furnace type, oven type etc.

- **Furnace:** (This can be either gas or electric), since the fuming from fluxes and metal vapors at high temperatures are so aggressive and volatile they have a tendency to attack the elements in an electric furnace, which can be very costly both in downtime and parts. Therefore it is recommended to use a gas fired furnace for reducing the sample mixture to a molten mass. The Carbon Monoxide produced by the flame also helps in reduction.

- **Assay oven or Kiln:** (Again this can be either Gas or electric), However the cupellation process must be done in a still air condition, with no drafts, therefore it is recommended for the small operation that this oven be electric.

- **Hand Tools / Hot Tools:** Most tools for the hot part of the assaying application are handmade they consist of tongs loaders and spatulas of many shapes, configurations, and sizes. This depending on such things as Top loading furnaces, Front loading Furnaces and ovens, distance needed between the assayer and hot work as well as controllability. Because of the strange configurations of these devices and lack of demand for them is the reason they are mostly handmade. A Pyrometer with a type 'K' Thermocouple should also be used to check temperatures if one is unsure about the temperature of the furnace

- **Miniature and Medical instruments:** An assayer will need such things as tweezers of different shapes and sizes, small enough to remove and manipulate some beads which can only be seen thru aid of magnification. Mostly by preference many of these can be found as Medical and Dental supplies. Magnifiers, Loupes, and a good microscope is a must. Optical comparators (Miniature), can also be used for determining an approximate bead weight thru size, using a mass scale.

- **Scales and accessories:** Generally a small operation must have three scales. The first must be capable of weighing larger masses up to 50Kg or 110 Lbs. These scales will generally cove receipt weight of any sample for testing. The second being a scale that will be used more than any other. This scale will be used in the range of 0 to 50 grams and measure in 10ths with a high degree of accuracy, used for measuring pulps and reagents.

The 3rd or next Scale, highly necessary is the 1mg to 50g scale this scale is slightly expensive depending on a new or used purchase and since it is designed for analytics and has no other purpose along with dependency on it being highly accurate. One must remember when dealing with all of the scales that it is highly necessary to calibrate all scales and the purchase and use of calibration weights are a must.

- **Sample splitter:** A sample splitter is an instrument which divides a primary sample into equal parts indicating a representative sample of the whole mass, the more times split the more accurate the representation. (Note there may still be an anomaly called nugget effect, in which a small particle of high grade material or metal may be in one pulp sample making the assay inaccurate, sometimes common in free milling ore, and can be eliminated by multiple samples of pulp being assayed).

- **Crushers and Pulverizers;** These are tools or machines necessary for comminution of the samples in preparation of pulps for an assay, it is desirable to have all material broken down into a pulp of less than 100-200M Tyler (149-74 microns) allowing good liberation and fracturing of the material so that it is more susceptible to particle collection during the Furnacing process. This breakdown is also preferential to most assayers. To do this the following is recommended

- **A Small laboratory Jaw Crusher:** To break down larger material for further processing and to fit in the smaller mills reducing time of comminution.

- **An Impact Mill or Disc pulverizer:** For dry materials

- **A Laboratory Ball mill or Tumbler with full charge of balls:** For wet grinding

Optional Tools / Safety Gear:

- **Classifier Screens:** An assayer may find it necessary to do a size by size analysis where it is necessary to see before or after comminution, where the most values lie. This is done by screening the materials, usually using specific ASTM Tyler mesh sizes to see what percentages of value are in a particular particle size, afterward doing multiple assays on those particular samples to determine processing methods. There are many methods to accomplish the above screening from hand screening to Ro-tap and other automatic screeners.

- **Safety Gear:** This such as protective fire retardant clothing, Face and Eye protection should always be worn by any individual performing any and all of the analysis of the Fire assay. Extreme temperatures capable of causing maiming and/or death are without doubt very probable without the proper personal protective equipment commonly referred to as PPE. Equipment such as respirators and breathing apparatuses are a necessity as metal fluxing and assay fumes are poisonous. Litharge, Lead oxide is considered a poison fatal if inhaled or ingested. Nitric acid fumes can also be considered very deadly. If you can find safety gear that you think will benefit you, buy it and use it, you cannot take too many precautions. (Examples of Gear demonstrated in class).

Procedures for a Fire Assay:

Please note, One very large and important step is for the assayer to keep extremely good redundant records, the next is total quality assurance by establishing a written procedure, updating and amending it periodically as needed, next are listed by function the steps in performing and living with a confident assay. Also note and understand that an assay no matter how large or small is only a representative of a sample accepted into your care.

One must also understand that a good assay carries no bias, and/or outside influence. The assayer must report exactly and only the seen results, any less or more is unacceptable. First part of the assaying is commonly referred to as firing, it requires a fireclay crucible capable of holding the sample and all of the reagents without boil over.

Startup

Prepare furnace and kiln start them up, so they may be brought up to temperature, usually 1850°F to 2000°F

- Gather all tools and safety equipment needed put them in their prospective positions for use.

- Move into sample area prepare sample by

- Weighing and Logging in description and exact weight of sample

- Clean or make sure all tools used for this assay are clean and free of debris (This is done as to never contaminate an assay whereby it would compromise its integrity.

- Comminution Crushing, grinding, and pulverizing of sample preparing for splitting (Wet samples may need to be dried prior).

- Splitting of sample to obtain a minimum of 2 samples of size and weight for firing. An example may be (2) separate 29.166 gram samples, so grams of material would be adequate for two assay samples in the case that they are needed or one of them fail.

- Bringing the assay to a readiness by taking a clean weight boat and weighing the proper assay weight before placing it in the crucible, in the case of doubling or tripling to get an adequate bead, do this 2 or 3 times.

- Now With a ceramic crayon mark and identify the crucibles with the identification of the sample.

As a General rule only, Use 3 times your pre-mixed reagents to each part of sample weight, mix thoroughly using a wooden or plastic mixing stick. Place crucibles in at temperature furnace for a minimum of 65 minutes. If using an unknown style and manufacturer of crucible be sure to charge it (Pre-heat) first, afterward removing it and checking for cracks and holes. The reason that pre-mixed flux (Reagents) were mentioned is that there may be certain alterations that might have to be made to tailor the flux to the ore, using a standard charting method.

Finishing Firing
- Check that mass is still in the crucible (No bubbling or boiling, if it is leave it in until it stops).
- Prepare conical mold by heating to insure there is no moisture present..
- Open Furnace, Remove crucible and pour into mold, one continuous pour rolling crucible to prevent dripping, (If lead spits from mold discard assay and perform a re-do.
- After cooling remove lead prill from mold, Knock off slag and pound lead into a square

Now you are ready for Cupellation!

Procedures for a Fire Assay (Cupellation): The next step of the assay is called cupellation or Cupelling the lead prill. Normal usage requires a concaved surface made of either Bone ash or Magnesite, both being made of a very porous material capable of absorbing the lead which will oxidize in this stage leaving only the precious metal bead, referred to as a Cupel. Cupellation at 1775°-1850°F takes approximately 5 minutes per gram of lead under good conditions.

1. Weigh Prill and record (This will let you know approximately when the cupellation is finished so as not to disturb the process).
2. Using a Scribe Identify the Unit or Cupel
3. Place the Squared lead prill in the cupel
4. Place the Cupel in the Kiln
5. Check on Cupellation about 5 minutes before due time to finish.
6. Remove Cupel from oven, allow to cool.
7. Remove bead and weigh recording weight.

❖ On a 29.166 gram assay, the bead weight in milligrams is equivalent ounces per ton of precious metal.

You are now ready to see how much of this bead is Gold and how much is Silver!

Identification and parting of a precious metal bead: There are Several ways to identify amounts of gold and silver in a PM bead.
   a) X-Ray Spectrometry (Very costly) Accurate
   b) Inductively coupled Plasma testing (ICP) Extremely Costly, Accurate
   c) Colorimetric Charting (Least costly, accuracy depends on individual doing analysis) non destructive

d) Dissolution using a heated 10% Nitric acid Solution dissolving all the silver in the bead. (Most accurate but requires a weighable amount of gold to determine value. (Inexpensive)

Parting using a heated 25% Nitric acid and 75% distilled water solution:
The following steps will direct you on use of Nitric, acid dissolving the Silver in the bead to determine the percentages of Au and Ag.

1. Prepare for use of a clean parting cup
2. Turn heat source on to a medium heat
3. Put on rubber gloves ,add parting solution to parting cup (filling cup half way) and add the bead with a set of tweezers
4. Using tweezers set parting cup on hot plate (will see steam after a few minutes)
5. Allow ample time for bead to stop boiling
6. Remove from heat make sure mixture is still pour off excess liquid.
7. Move (Wash) the remaining material with distilled water.
8. Wash again with household ammonia
9. Wash a 3rd time with distilled water
10. Wash into a scorifier dish place a propane mini torch heating the mixture to anneal it to its natural color of gold. Wait for cooling
11. Place into a clean weight boat, weigh and calculate percentage of precious metal bead.
12. Record results.

Parting takes roughly 5 minutes to complete.

It is strongly recommended when suspicion of the Platinum group is present in an assay bead, usually indicated by a dull gray bead which has not fused, a silver bead that will not change when parted that the beads be sent out or finished by machine. There are many methods that can accurately determine the amounts of the platinum group with the benefit of also reporting other elements. Two of these methods are the

Spectrographic analysis, very accurate yet inexpensive. The next is the newer Inductively Coupled Plasma mass spectrometry analysis (ICP).

Assaying for the Platinum Group by Fire with a machine finish, (**Spectrographic analysis**)

The Metals of the Platinum Group, Platinum, Palladium, Rhodium, Ruthenium, Iridium, can normally be determined by spectrographic analysis of a fire-assay bead which appears Silver, Gray or dull with an erratic surface, in Color. Before this point the procedures for the fire assay are as described in the normal assaying of gold and silver. Spectrographic analysis is done by placing fire-assay bead into a graphite electrode containing Osmium ICP Standard reagent and arced in an argon-oxygen atmosphere, whereby the scale of metals can be read and reported. This method does require the high tech equipment and is probably cost prohibitive to the small mining operation or operator.

Assaying for Gold by Fire with a machine finish, (**AA, or Atomic Absorption**)

This is accomplished by dissolution of the entire Silver/Gold dore bead in aqua regia determining the gold content by AA (Atomic Absorption). AA is a very instrumental method of determining element concentration by introducing an element in its atomic form to a light beam of an appropriate wavelength causing the atom to absorb light. The reduction in the intensity of the light beam directly correlates with the concentration of the elemental atomic species.

Hoffman, E.L. , Clark, J.R. and Yeager, J.R. 1998. Gold analysis - Fire Assaying and alternative methods. Exploration and Mining Geology, Volume 7, pp. 155-160.

Above are just several methods of common assays by laboratories but all of them being time proven with accuracy depending on the operational ability of the machine coupled with the dependability, accuracy and education of the technician or assayer performing the analysis.

The Next Pages you will see different examples and explanations of the assaying compounds, what they do and how they work for different type ore charges. Ore and Assay charges do can and do vary from host to host type rock. Remember how they work when attempting to adjust your reagents (Tailoring) charges to a particular type ore where a standard flux fails to produce results.

After pointing out some of these, as a final discussion we will address slag's and what the colors of that glass means.

Assay Fluxes the Composition.

Common Standard assay fluxes, as stated before are proven charges developed by a chemist using both complicated sciences building a value of material which will collectively produce good results by capturing all of the values of precious metals in a sample.

As a repeat of the mixture for a **standard assay flux**, the composition is as follows;

69% Litharge, Lead oxide (PbO)

23.3% Soda ash, Sodium Carbonate (Na2CO3)

Anhydrous Borax 3.5%. (N2B4O7)

 Silica Sand minimum size 70 Mesh 2.2% (SiO2).

 Wheat Flour 2% (Carbon Source).

General Use of this Flux will produce a Lead prill of at least 30 grams which is considered to be an ample amount of lead for collection of the precious metals using a 3 parts flux to 1 part sample ratio by weight. (NOT BY VOLUME).

Occasionally where directly assaying sulfide materials, one may choose not to use flour in the compound since sulfides (depending on the amounts have a good reduction capability. This is an adjustment to be determined by the assayer.

When dealing with sulfides as a norm the addition of Potassium Nitrate is added as an oxidizer to turn the metals from within, from sulfides to oxides which readily combine with the lead precipitates (collection) carrying the whole lot to the bottom of the crucible.

Characteristics of the dry Chemicals

Fusion in general normally takes place in the furnace typically between 2,200°F and 2,500°F Assay fusion can be done efficiently using a broad range of temperatures from 1850°F to 2000°F depending on the ore composition. But with the composition of fluxing reagents materials that normally melt or fuse at a higher temperature are adjusted by the reagents to create lower fusion and melting temperatures.

Most of the dry chemicals used in the fire assay have different appearances characteristics and do different things.

The efficiency of separation and collection all depend on the quality of the slag material that is formed by the dry chemicals.

For an example a molten slag which has and extremely high viscosity may not allow the drop of collected metal to the bottom of the crucible in the allotted time in the furnace (Normally 1 to 1.5 hours at temperature), suspending value in the slag.

Silica normally forms the bases of the fluxing and has the capability of dissolving most of the metal oxides. Silica also has a high melting point of 3,133°F and has tendencies of forming highly viscous slags.

Anhydrous borax has a much lower melting point of approximately 1472°F and is highly acidic at this temperature aiding in dissolving of the base metals keeping them suspended while the reactions take place for collection. The Borax works hand in hand with the Silica to accomplish this reaction.

Sodium Carbonate has been used for a very long time in fluxes, it not only increases the clarity in slags but also decreases viscosity, and the large benefit of this material is by reducing the viscosity it aids in the prevention of entrainment.

Calcium Fluoride is extremely beneficial in reducing viscosity by replacing silica with fluoride ions.

Potassium Nitrate often referred to as a Nitre, beneficial for supplying high volumes of oxygen turning sulfides into oxides in a expeditious manner of reaction in the fusion state. Care in over addition must be taken when using this reagent as they can also be the culprit of loss by volatizing of some gold and silver.

Manganese Dioxide this addition is beneficial when trying to oxidize a sum of base metals; it works well in cases of heavier Zinc, and when not wanting to use the Niters.

Litharge, Lead Oxide is the primary character in the assay flux mixture. The small particles of lead oxide, assisted by the carbon that the flour or other reducer creates, produces tiny droplets of pure lead metal, raining thru the mixed slags collecting all of the precious metal within. The droplets all rain thru the material having a higher specific gravity assembling at the base of the crucible alloyed as a lead PM combination waiting to be poured into a mold for further processing. The right amount of litharge added will create a lead prill equaling at least 30 grams of metal in a one ton fire assay. Any less than 30 grams would be questionable that there were enough of the metal oxides to collect and do a complete job.

There are many other materials which can be added to a fire assay to do a better job on certain materials.

Generally speaking a professional opinion that has been expressed many times by many assayers feel that anyone assaying should do or have a broad spectrum analysis done to find out just what's in the rock as far as other minerals go, to insure a proper flux mixture is made and see what might interfere with providing a good assay. Or to see what other contaminates such as High amounts of Nickel, Arsenic, Sulfur, Cobalt or any other numbers of material which may affect the proper reduction fusion or cupellation that may be included which may throw an assay off accuracy.

It is extremely important that it is understood that any number of large amounts of contaminates can present an inaccurate assay, and unless the person performing the assay is a seasoned assayer that in the case that one has to report his/her assays another assayer or umpire assayer should be consulted or used to verify results in a controlled sample.

If assaying for oneself there may not need the necessity to verify or produce an extremely accurate assay or in the case that the person doing the assay wants to just get a general idea of the values present. However it is still highly recommended to still get a broad spectrum analysis to know what you are dealing with.

Assaying of Specific Ore Composites.

The Assays

Sulfide-Sulphide and Pyritic Ores

These ores containing larger proportions of Sulfides for example, Pyrite, Chalcopyrite, Arsenopyrite, along with other Gold and Silver bearing sulfides require a little special treatment using Nitre or a few other methods to rid or oxidize the ore of its sulfur content, producing an oxide metal. This as stated above can be done in several ways to obtain a good assay and collection of the precious metals.

One may first roast the ore at a temperature of approximately 1,000°F with a good supply of air for oxidation for a minimum of 1 hour or until there is no presence of smoking during the roast, after which a standard assay flux may be used to complete the assay fusion.

Using a the addition, Potassium Nitrate, to the standard assay flux 90 grams per a one ton assay, and which is assumedly pre-mixed and making the addition of 18 to 20 grams of nitre will give this mixture the oxidizing power it needs during the furnacing to perform an assay without pre treatment of the ore (Roasting). One must take care not to over add the nitre as it may cause a lot of foaming and a boilover resulting in loss of the assay, not to mention what it does to the furnace lining. This method works best when a trial run is done to make sure that all material has properly dissolved into the slag. Any Matte that may be on top of the lead prill is an indicator that more nitre must be used to accomplish the full reaction of the additional reagent.

Another great method is the well known Iron nail assay whereby in the standard one ton assay the addition of 4 to 5, 8d nails are added with the ore and the standard 90 grams of flux (5 to 6 is best if ore is unknown) if too many nails are added the remaining portion of nails not dissolved have to be removed before pouring, this age old method uses the iron to desulferize the material contained within.

Intentionally Blank

The Heavy Copper bearing minerals.

In an assay of this type of ore the heavy amounts of copper (a base metal) has a great tendency to collect in larger quantities along with the gold and silver and therefore carried to the next step which is cupellation.

As this happens unless properly treated this particular material can cause losses of both gold and silver resulting in an extremely inaccurate assay. This material requires the addition of a scorification process.

The process for a one assay ton test is as follows;

1. Instead of the standard 90 grams of assay flux the addition of another 30 grams must be used equaling 120 grams

2. Furnace at approximately 1,850°F for one hour or until fusion is still.

3. Pour into mold, cool, knock off slag. A lead prill of approximately 40 grams will be produced, requiring the use of at least a 3" scorifier dish for the next step.

4. Place the prill in the scorifier dish, using a 50/50 mixture of Silica and Anhydrous Borax

5. Furnace the material at approximately 1,850°F to 1,950°F until the lead begins to slag over and pour, adding 50 grams of clean lead each time this is repeated

6. Duplicate this process as many times as it takes to soften the lead before Cupelling, once the lead is no longer hard most of the copper contaminate has been removed

7. At this point when the lead is no longer hard it may be cupelled.

The previous 2 pages show a few but the most common processes that someone new to assaying might encounter.

This book was presented to primarily introduce layman or persons interested in the art and science of the fire assay

Thru time, experiences, much research and practice one may become an assayer and expert of the art.

A metallurgical practice of the ages, Assaying by Fire

Disclaimer:

My Name is N.E. Kendall I am a mining and processing consultant, by presenting the following information to you, it is my intention to only assist and provide information with mining facts, opinions and processes practiced today by, Assayers and myself.—Some of this information may contain some unintentional errors and is not meant to replace proven methods, laws or procedures—I am not a Lawyer---All stated laws should be thoroughly proven before followed by an attorney. All opinions are of my own and stated by my own personal beliefs, cited in this book as a first amendment right to the US constitution. Some of this information may no longer be valid as it is subject to change without notice—Common practices and processes may change with technology and may include differences not stated or may not work on certain ores. It is the sole responsibility of the person reading this book to investigate the processes and instructions before using them. There may be typos included in these writings as well. I make no guarantees as to the safety, practice, or outcome of any which have been stated.